Units and conversion charts

a handbook
for engineers and scientists

Units and conversion charts

a handbook
for engineers and scientists

Theodore Wildi

President, Sperika Enterprises Ltd.

Professor Emeritus of Laval University
Department of Electrical Engineering

 **IEEE
PRESS**

The Institute of Electrical and Electronics Engineers, Inc., New York

ISBN 0-87942-273-4

IEEE Order Number: PP0267-5

For information about our audio products, write us at:
Newbridge Book Clubs, 3000 Cindel Drive, Delran, NJ 08370

Printed in the United States of America
10 9 8 7 6 5 4 3 2 1

PREFACE

As systems of measurement evolved in various parts of the world, they produced a large number of units. This handbook shows in an exceptionally clear and useful way how these units are related to each other, and how they are defined. Its principal advantage lies in the conversion of units, a process made exceedingly simple by a set of new conversion charts. They enable the engineer, scientist and technician to make rapid and clear-cut conversions between units of the American Customary system, the English system, former metric systems, and the International System of Units (SI).

The conversion charts rank the units by order of size so that the relationship between any two units can be found quickly and without ambiguity. They significantly reduce the time usually needed to consult handbooks, tables and so forth, in solving engineering and scientific problems.

The information contained herein is based upon the latest data on units and quantities published by the American National Standards Institute, The Institute of Electrical and Electronics Engineers, the Bureau International des Poids et Mesures, the Committee on Data for Science and Technology (CODATA) and by the International Organization for Standardization.

The conversion methodology was first devised by the author in the early 1970's, and since then thousands of engineers and scientists have discovered its usefulness. The methodology was explained in two pioneering books "Units" and "Understanding Units" which are now out of print, having been replaced by this current up-dated book.

I want to express my thanks to the many individuals who offered their helpful comments. In particular, I want to mention Dr. Gilles Y. Delisle, Dr. Hugh Preston-Thomas, Mr. Guy W.-Richard, Dr. Jacques Vanier and Mr. K. C. Ford.

Last, but not least, I thank Lucie Veilleux for typing the manuscript and putting it into computerized form, and my son Karl for drawing the charts.

T.W.

CONTENTS

Introducing The Conversion Charts

Methodology

In this book, the units of various quantities such as force, pressure, viscosity, etc., are displayed by means of conversion charts.

The charts show the relative size of a unit by the position it occupies on the page. The largest unit is at the top, the smallest at the bottom and intermediate units are ranked in between.

The units are connected by arrows, each of which bears a number. This number is the ratio of the larger to the smaller of the units that are connected, and hence its value is always greater than one. The arrow always points toward the smaller of the two units.

In Figure 1, for example, six units of length - the mile, meter, yard, foot, inch and millimeter - are positioned in descending order of size*. The numbers show the relative magnitude of the connected units; the yard is 3 times larger than the foot, the foot is 12 times larger than the inch, and so on. With this arrangement we can convert any UNIT A into any UNIT B by the following simple procedure. Starting from UNIT A we follow any path that ends up at UNIT B while observing the rule:

with the arrow   MULTIPLY

against the arrow   DIVIDE

Because the arrows point downwards, we multiply when moving down the chart, and divide when moving up.

* In most English-speaking countries, including Canada, the SI unit of length is spelled metre, and a unit of volume is spelled litre. On the other hand, the spelling of these two units in the United States is usually meter and liter. In this book, we have adopted the meter, liter spelling.

Example 1:
Convert 2.5 feet to millimeters.

Solution:
Starting from **foot** and moving towards **millimeter** we always move in the direction of the arrows. We must therefore *multiply* the numbers associated with each line:

$$2.5 \text{ feet} = 2.5 \, (\times \, 12) \, (\times \, 25.4) \text{ millimeters}$$
$$= 762 \text{ millimeters}$$

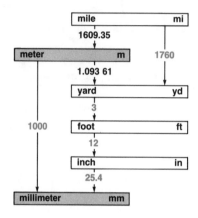

Figure 1

Example 2:
Convert 95 040 inches into miles.

Solution:
Starting from **inch** and moving toward **mile**, we always move against the direction of the arrows. We must therefore *divide* by the number associated with each line:

$$95 \, 040 \text{ inches} = 95 \, 040 \, (\div \, 12) \, (\div \, 3) \, (\div \, 1760) \text{ miles}$$

$$= \frac{95 \, 040}{12 \times 3 \times 1760} \text{ miles}$$

$$= 1.5 \text{ miles}$$

Flyers

When units are far apart, several multiplications or divisions must be made to convert from one unit to another. To reduce the arithmetic, "flyers" are introduced to bypass a series of units. A flyer also bears a number, whose value is equal to the product of the numbers which the flyer bypasses. In Figure 1, for example, the flyer 1000 joining meter-millimeter is equal to the product of the numbers 1.093 61 $\times$ 3 $\times$ 12 $\times$ 25.4. (1.093 61 represents 6-figure accuracy).

Flyers provide additional paths between units, and any of these paths may be followed without affecting the conversion result.

Example 3:

Convert 750 inches into meters.

Solution a:

750 inches $=$ 750 ($\div$ 12) ($\div$ 3) ($\div$ 1.093 61)
$=$ 19.05 meters

Solution b:

750 inches $=$ 750 ($\times$ 25.4) ($\div$ 1000)
$=$ 19.05 meters

note that solution b using the flyer is quicker because it involves only one multiplication (with the arrow) and one division (against the arrow).

SI units

SI units and their multiples and submultiples appear in *red* boxes on the conversion charts. Because they are connected by flyers that are multiples of ten, it is possible to move swiftly from one end of a chart to the other should a conversion between widely-separated units be required. American customary units and non-SI metric units can readily be converted to SI or vice-versa by the technique we have just described.

Symbols

The symbols for all SI units (and for many non-SI units) appear on the right-hand side of each box. Symbols have been omitted whenever a clear and authoritative consensus was lacking.

Numerical Values

Numerical values between two units are shown either in red or in black.

RED numbers are exact, by definition.

BLACK numbers are accurate to the number of significant figures shown.

A useful feature of the charts is that exact conversion accuracy between units can be obtained by following paths that involve red numbers. These paths are usually erratic, and may require one or more multiplications and divisions.

For example, referring to the chart on LENGTH, page 25, we find that 1 meter = 39.3701 inches (6-figure accuracy). But if we follow a path involving red numbers we find that 1 meter = 1 ($\times$ 1000) ($\div$ 25.4) = 39.370 078... inches *exactly*.

Converting to SI Base Units

Any derived unit of the SI can be expressed in terms of one or more of the seven SI base units. This "base dimension" is displayed in brackets on the left-hand side of each conversion chart. For example, referring to the chart on ENERGY, page 19, we see that the SI base dimension of the joule is kg·m^2/s^2. This relationship is more than one of simple equivalence: 1 joule is *exactly* equal to 1 kg·m^2/s^2. Appendix V describes how the dimension of a unit is established.

Multiples and submultiples

Multiples and submultiples of SI units carry prefixes according to Table 1. Using this prefix designation, a length of 1000 meters is called a *kilo*meter, while 1/1000 of a meter (0.001 meter) is called a *milli*meter. Similarly, 1/10 of a liter is called one *deci*liter and 100 liters is one *hecto*liter.

However, in the interest of simplicity, and to minimize the possibilities of error, the prefixes centi, deci, deca and deci are often avoided. For example, builders and machinery manufacturers commonly quote lengths only in meters and millimeters.

Writing Numbers

To simplify the reading of numbers having many digits, the digits are set in groups of three, separated by a space. The spaces are counted to the left and to the right of the decimal marker. This style eliminates the need for commas as group separators. For example, we write

13 754 instead of 13,754

25 146.241 32 instead of 25,146.24132

From an international standpoint, this group-of-three rule also prevents confusion because many countries use the comma as a decimal marker.

Handling special conversions

In solving some engineering problems, the answer may contain several unusual units that seem quite unrelated to the solution we are seeking. Such units should be first expressed in terms of SI base units. The new units can then be simplified to yield the result we are seeking.

Example 4:
In braking a train, it was found that the amount of energy consumed was

$$\text{energy} = 75 \times 10^7 \ \frac{\text{ton} \cdot \text{yard} \cdot \text{mile}}{\text{hour} \cdot \text{minute}}$$

We want to express this unusual result in terms of Btu.

TABLE 1   SI PREFIXES AND THEIR SYMBOLS

Multiplier	Exponent Form	Prefix	SI Symbol
1 000 000 000 000 000 000	10^{18}	exa	E
1 000 000 000 000 000	10^{15}	peta	P
1 000 000 000 000	10^{12}	tera	T
1 000 000 000	10^{9}	giga	G
1 000 000	10^{6}	mega	M
1 000	10^{3}	kilo	k
100	10^{2}	hecto	h
10	10^{1}	deca	da
0.1	10^{-1}	deci	d
0.01	10^{-2}	centi	c
0.001	10^{-3}	milli	m
0.000 001	10^{-6}	micro	μ
0.000 000 001	10^{-9}	nano	n
0.000 000 000 001	10^{-12}	pico	p
0.000 000 000 000 001	10^{-15}	femto	f
0.000 000 000 000 000 001	10^{-18}	atto	a

Solution:

Letting {H} be the number of Btu, we write the equation

$$\{H\}\ \text{Btu} = 75 \times 10^7\ \frac{\text{ton} \cdot \text{yard} \cdot \text{mile}}{\text{hour} \cdot \text{minute}}$$

thus
$$\{H\} = 75 \times 10^7\ \frac{\text{ton} \cdot \text{yard} \cdot \text{mile}}{\text{hour} \cdot \text{minute} \cdot \text{Btu}}$$

Next, using the charts, we convert each unit to SI base units, as follows:

1 ton $= 1\ (\div\ 1.102\ 31)\ (\times\ 1000)$ kg $= 907.186$ kg   (see p. 30)

1 yard $= 1\ (\div\ 1.093\ 61)$ m $= 0.914\ 40$ m   (see p. 25)

1 mile $= 1\ (\times\ 1.609\ 344)\ (\times\ 1000)$ m $= 1609.34$ m

1 hour $= 3600$ s

1 minute $= 60$ s

1 Btu $= 1\ (\times\ 1.054\ 35)\ (\times\ 1000)$ J $= 1\ 054.35$ kg·m²·s⁻² (see p. 19)

Note that Btu is converted into SI base units by using the dimension of the joule, displayed in brackets on the left-hand side of the ENERGY chart.

Substituing these values in the above equation, we obtain

$$\{H\} = 75 \times 10^7\ \frac{(907.186\ \text{kg}) \times (0.9144\ \text{m}) \times (1609.34\ \text{m})}{(3600\ \text{s}) \times (60\ \text{s}) \times (1054.35\ \text{kg·m}^2\text{·s}^{-2})}$$

$$= 4.396 \times 10^6$$

Note that all units cancel, showing that the original answer did indeed have the dimension of energy. The energy consumed is therefore 4.396×10^6 Btu.

Conversion Charts

SI UNITS: RED BOXES

ALL OTHER UNITS: WHITE BOXES

RED NUMBERS are exact, by definition

BLACK NUMBERS are accurate to the number
of significant figures shown

Conversion Rule

WITH THE ARROW - MULTIPLY

AGAINST THE ARROW - DIVIDE

SI PREFIXES
SYMBOLS AND VALUES

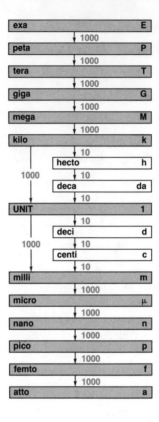

exa	E

↓ 1000

peta	P

↓ 1000

tera	T

↓ 1000

giga	G

↓ 1000

mega	M

↓ 1000

kilo	k

↓ 10

hecto	h

1000 ↓ 10

deca	da

↓ 10

UNIT	1

↓ 10

deci	d

1000 ↓ 10

centi	c

↓ 10

milli	m

↓ 1000

micro	μ

↓ 1000

nano	n

↓ 1000

pico	p

↓ 1000

femto	f

↓ 1000

atto	a

ACCELERATION

ANGLE

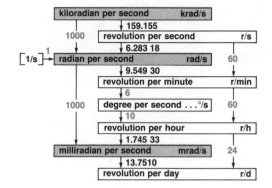

Example: Convert 1800 r/min to radians per second

Solution: 1800 r/min = 1800 (÷ 9.549 30) rad/s = 188.495 rad/s

ANGULAR FREQUENCY
CIRCULAR FREQUENCY
PULSATANCE

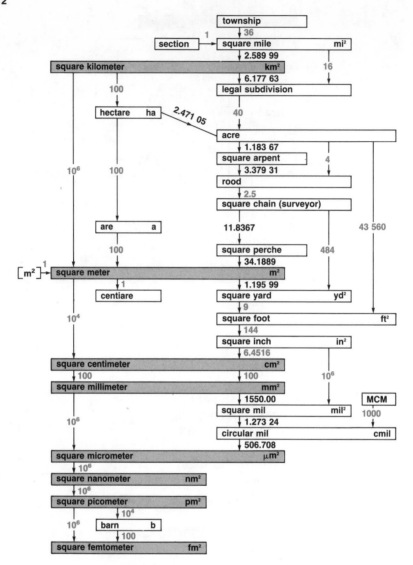

Example: Convert 4 square miles to hectares

Solution: 4 mi² = 4 (× 2.589 99) (× 100) ha = 1036 hectares

CONCENTRATION

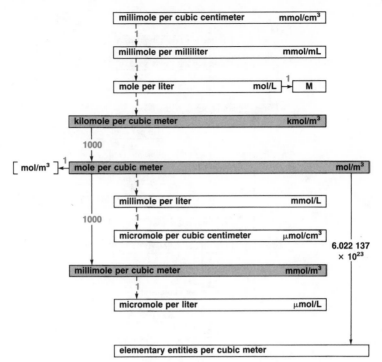

| millimole per cubic centimeter | mmol/cm³ |

| millimole per milliliter | mmol/mL |

| mole per liter | mol/L | → | M |

| kilomole per cubic meter | kmol/m³ |

1000

[mol/m³] ← | mole per cubic meter | mol/m³ |

| millimole per liter | mmol/L |

1000

| micromole per cubic centimeter | μmol/cm³ |

6.022 137
× 10²³

| millimole per cubic meter | mmol/m³ |

| micromole per liter | μmol/L |

| elementary entities per cubic meter |

DENSITY (mass per unit volume)

Example: Convert 60 lb per cubic foot to tons per cubic yard
Solution: 60 lb/ft³ = 60 (× 16.0185) (÷ 1000) (÷ 1.186 55) = 0.81 ton/yd³

ELECTRICITY

RESISTANCE

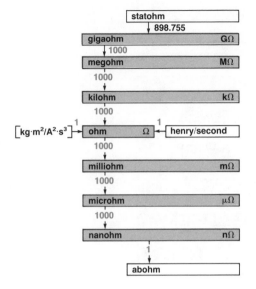

statohm	
↓ 898.755	
gigaohm	GΩ
↓ 1000	
megohm	MΩ
↓ 1000	
kilohm	kΩ
↓ 1000	

$[kg·m^2/A^2·s^3]$ → ohm Ω ← henry/second

↓ 1000	
milliohm	mΩ
↓ 1000	
microhm	μΩ
↓ 1000	
nanohm	nΩ
↓ 1	
abohm	

CONDUCTANCE

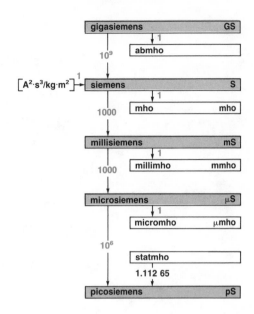

gigasiemens	GS
↓ 1	
abmho	
10⁹	

$[A^2·s^3/kg·m^2]$ → siemens S

↓ 1	
mho	mho
↓ 1000	
millisiemens	mS
↓ 1	
millimho	mmho
↓ 1000	
microsiemens	μS
↓ 1	
micromho	μmho
↓ 10⁶	
statmho	
↓ 1.112 65	
picosiemens	pS

INDUCTANCE

stathenry	
↓ 8.987 55 × 10¹¹	

$[kg·m^2/A^2·s^2]$ → henry H ← ohm second

↓ 1000	
millihenry	mH
↓ 1000	
microhenry	μH
↓ 1000	
nanohenry	nH
↓ 1	
abhenry	

CAPACITANCE

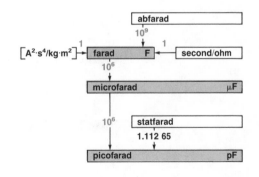

abfarad	
↓ 10⁹	

$[A^2·s^4/kg·m^2]$ → farad F ← second/ohm

↓ 10⁶	
microfarad	μF
↓ 10⁶	
statfarad	
↓ 1.112 65	
picofarad	pF

CURRENT DENSITY

ELECTRIC FIELD STRENGTH

ELECTRICITY

CURRENT

CHARGE

ELECTROMOTIVE FORCE

Example: Express 3 kilovolts in terms of SI base dimensions

Solution: 3 kV = 3 (× 1000) (× 1) kg·m²/(A·s³) = 3000 kg·m²/(A·s³)

ELECTRICITY

MASS RESISTIVITY

VOLUME RESISTIVITY

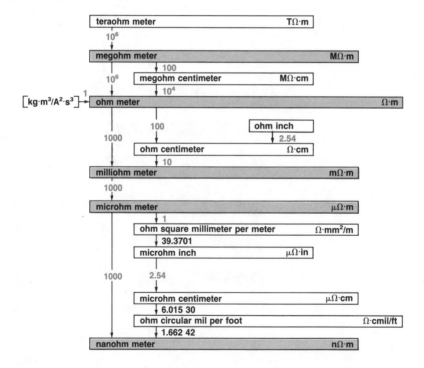

Example: Convert 60 ohm circular mil per foot to ohm cm

Solution: 60 Ωcmil/ft = 60 ($\times$ 1.662 42) ($\div$ 1000) ($\div$ 1000) ($\div$ 10) = 9.974 52 $\times$ 10^{-6} Ω·cm

ENERGY

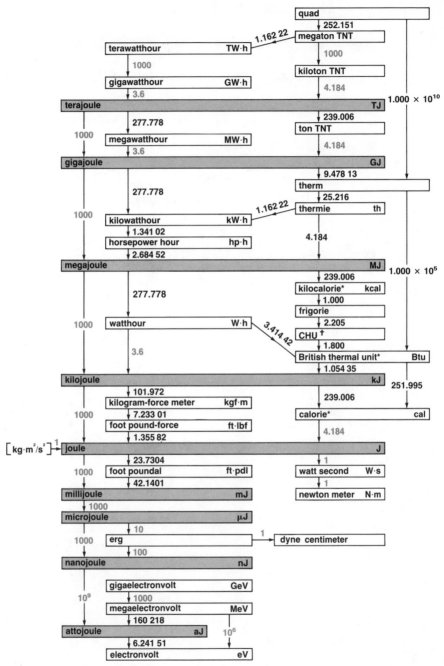

Note: Btu and calorie are *thermochemical* units

CHU is the heat required to raise the temperature of 1 lb of water by 1°C

ENERGY (calorie units)

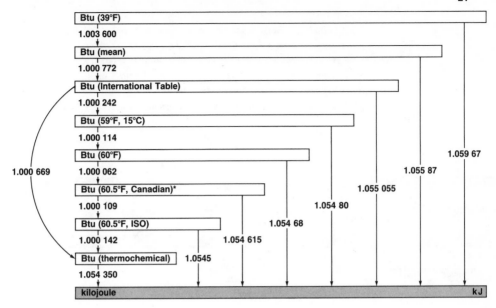

Example: Convert 4000 Btu(60°F) to Btu (thermochemical)

Solution: 4000 Btu(60°F) = 4000 (× 1.054 68) (÷ 1.054 350) Btu(thermochemical) = 4001.25 Btu(thermochemical)

ENERGY (atomic and physical relationships)

c = velocity of light = 2.997 924 58 × 10⁸ m/s = {c} m/s

k = Boltzmann constant = 1.380 658 × 10⁻²³ J/K = {k} J/K

h = Planck constant = 6.626 0755 × 10⁻³⁴ J·s = {h} J·s

c_2 = second radiation constant = 1.438 769 × 10⁻²m·K = {c_2} m·K

N_A = Avogadro constant = 6.022 1367 × 10²³ mol⁻¹ = {N_A} mol⁻¹

{R_∞} = Rydberg constant = 1.097 3731 × 10⁷ m⁻¹ = {R_∞} m⁻¹

{ } = numerical value of constant when its dimension is expressed in SI units

cal_{th} = thermochemical calorie

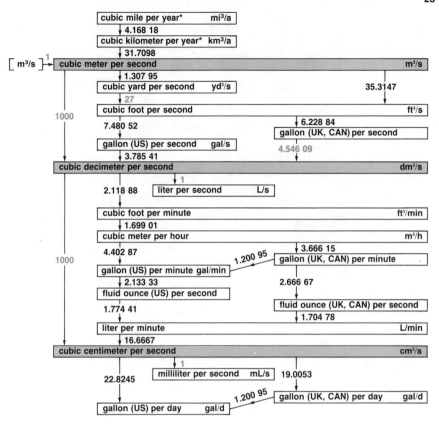

cubic mile per year* mi³/a
↓ 4.168 18
cubic kilometer per year* km³/a
↓ 31.7098
[m³/s]─1── cubic meter per second m³/s
↓ 1.307 95
cubic yard per second yd³/s   35.3147
↓ 27
cubic foot per second ft³/s
7.480 52   ↓ 6.228 84
gallon (UK, CAN) per second
gallon (US) per second gal/s
4.546 09
↓ 3.785 41
cubic decimeter per second dm³/s
1000   ↓ 1
2.118 88   liter per second L/s
↓
cubic foot per minute ft³/min
↓ 1.699 01
cubic meter per hour m³/h
4.402 87   ↓ 3.666 15
1.200 95   gallon (UK, CAN) per minute
gallon (US) per minute gal/min
↓ 2.133 33   2.666 67
fluid ounce (US) per second
1.774 41   fluid ounce (UK, CAN) per second
↓ 1.704 78
liter per minute L/min
↓ 16.6667
cubic centimeter per second cm³/s
22.8245   ↓ 1   19.0053
milliliter per second mL/s
1.200 95   gallon (UK, CAN) per day gal/d
gallon (US) per day gal/d

*year = 365 days.

Example: Convert 250 gallons(US) per minute to liters per second
Solution: 250 gal(U.S.)/min = 250 (× 2.133 33)(× 1.774 41)(× 16.6667)(÷ 1000) L/s = 15.7725 L/s

FORCE

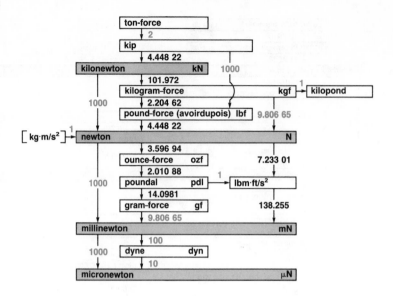

Example: Convert 30 lbf to newtons

Solution: 30 lbf = 30 (× 4.448 22) N = 133.447 N

FREQUENCY

LENGTH

Example: Express the Bohr first electron radius (a_o) in fermis

Solution: a_o = 1 ($\times$ 21.8100) ($\times$ 2.426 31) ($\times$ 1000) fermi = 52 917.8 fermi

LIGHT

LUMINANCE

ILLUMINANCE

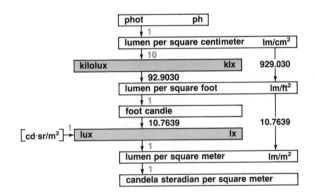

LUMINOUS INTENSITY

LUMINOUS FLUX

MAGNETIC MOMENT (atomic units and quantities)

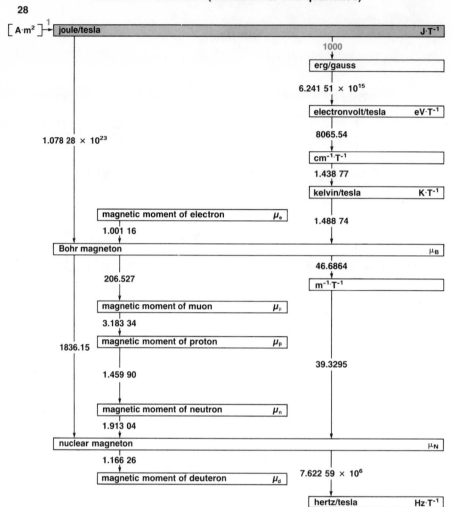

$[A·m^2]$ —1→ **joule/tesla** ————————————————————————— $J·T^{-1}$

1000
↓
erg/gauss

$6.241\ 51 \times 10^{15}$
↓
electronvolt/tesla — $eV·T^{-1}$

8065.54
↓
$cm^{-1}·T^{-1}$

1.438 77
↓
kelvin/tesla — $K·T^{-1}$

$1.078\ 28 \times 10^{23}$

1.488 74

magnetic moment of electron — μ_e

1.001 16
↓
Bohr magneton — μ_B

46.6864
↓
$m^{-1}·T^{-1}$

206.527
↓
magnetic moment of muon — μ_μ

3.183 34
↓
magnetic moment of proton — μ_p

1836.15

1.459 90
↓
39.3295

magnetic moment of neutron — μ_n

1.913 04
↓
nuclear magneton — μ_N

1.166 26
↓
magnetic moment of deuteron — μ_d

$7.622\ 59 \times 10^6$
↓
hertz/tesla — $Hz·T^{-1}$

MAGNETISM

MAGNETOMOTIVE FORCE

MAGNETIC FLUX

MAGNETIC FIELD STRENGTH

MAGNETIC FLUX DENSITY

MASS

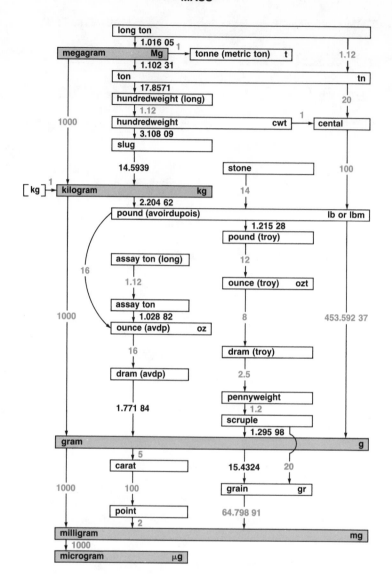

Example: Convert 2 kg to troy ounces **exactly**

Solution: For exact conversions we follow the arrows bearing red numbers. Thus:

$$2\,kg = 2(\times 1000)(\times 1000)(\div 64.798\,91)(\div 20)(\div 1.2)(\div 2.5)(\div 8)\,ozt = 64.301\,49\ldots\ldots ozt$$

MASS (atomic units and quantities)

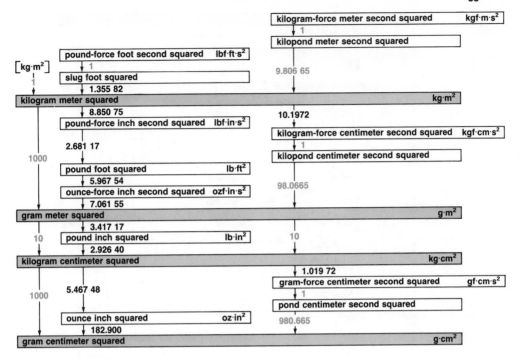

Example: Convert 80 lb·ft² into kg·m²
Solution: 80 lb·ft² = 80 (× 5.967 54) (× 7.061 55) (÷ 1000) kg·m² = 3.371 21 kg·m²

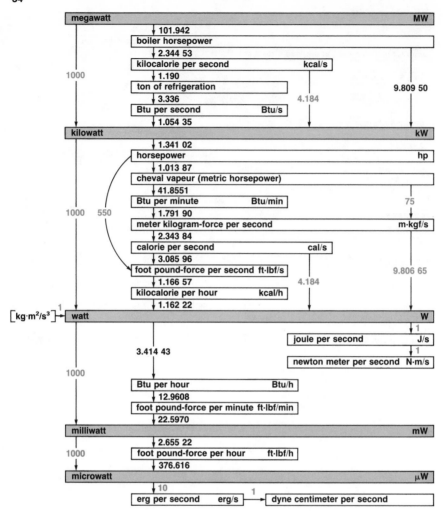

megawatt — MW
↓ 101.942
boiler horsepower
↓ 2.344 53
kilocalorie per second — kcal/s
↓ 1.190
ton of refrigeration
↓ 3.336
Btu per second — Btu/s
↓ 1.054 35
1000 — 4.184 — 9.809 50

kilowatt — kW
↓ 1.341 02
horsepower — hp
↓ 1.013 87
cheval vapeur (metric horsepower)
↓ 41.8551
Btu per minute — Btu/min
↓ 1.791 90
meter kilogram-force per second — m·kgf/s
↓ 2.343 84
calorie per second — cal/s
↓ 3.085 96
foot pound-force per second ft·lbf/s
↓ 1.166 57
kilocalorie per hour — kcal/h
↓ 1.162 22
1000 — 550 — 75 — 4.184 — 9.806 65

$\left[\text{kg·m}^2/\text{s}^3\right]$ 1 → watt — W
↓ 1
joule per second — J/s
↓ 1
newton meter per second — N·m/s

3.414 43
1000
Btu per hour — Btu/h
↓ 12.9608
foot pound-force per minute ft·lbf/min
↓ 22.5970

milliwatt — mW
↓ 2.655 22
1000
foot pound-force per hour — ft·lbf/h
↓ 376.616

microwatt — μW
↓ 10
erg per second — erg/s —1→ dyne centimeter per second

Note: Btu and calorie are *thermochemical* units.

Example: Convert 12 boiler horsepower into kilowatts
Solution: 12 boiler hp = 12 (÷ 101.942) (× 1000) kW = 117.714 kW

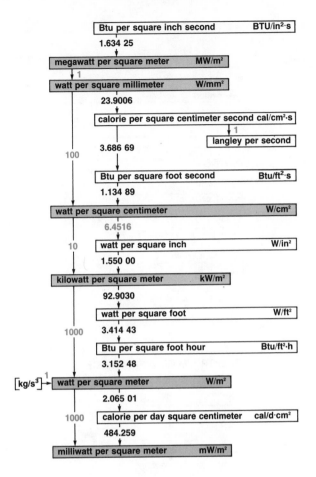

Btu per square inch second	BTU/in²·s

1.634 25

megawatt per square meter	MW/m²

1

watt per square millimeter	W/mm²

23.9006

calorie per square centimeter second cal/cm²·s	

1

langley per second	

3.686 69

100

Btu per square foot second	Btu/ft²·s

1.134 89

watt per square centimeter	W/cm²

6.4516

watt per square inch	W/in²

10

1.550 00

kilowatt per square meter	kW/m²

92.9030

watt per square foot	W/ft²

1000   3.414 43

Btu per square foot hour	Btu/ft²·h

3.152 48

$\left[kg/s^3\right]$ 1

watt per square meter	W/m²

2.065 01

1000

calorie per day square centimeter	cal/d·cm²

484.259

milliwatt per square meter	mW/m²

Note: Btu and calorie are *thermochemical* units

POWER (horsepower units)

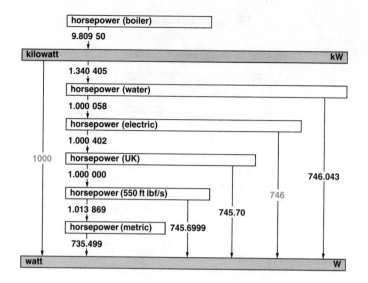

horsepower (boiler)

9.809 50

kilowatt **kW**

1.340 405

horsepower (water)

1.000 058

horsepower (electric)

1.000 402

1000

horsepower (UK)

1.000 000

horsepower (550 ft lbf/s)

1.013 869

horsepower (metric)   745.6999

735.499

746.043

746

745.70

watt **W**

POWER LEVEL DIFFERENCE

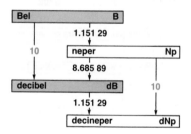

Bel **B**

1.151 29

10

neper **Np**

8.685 89

decibel **dB**   10

1.151 29

decineper **dNp**

Example: Convert 80 decibels to nepers
Solution: 80 dB = 80 (÷ 8.685 89)Np = 9.21 Np

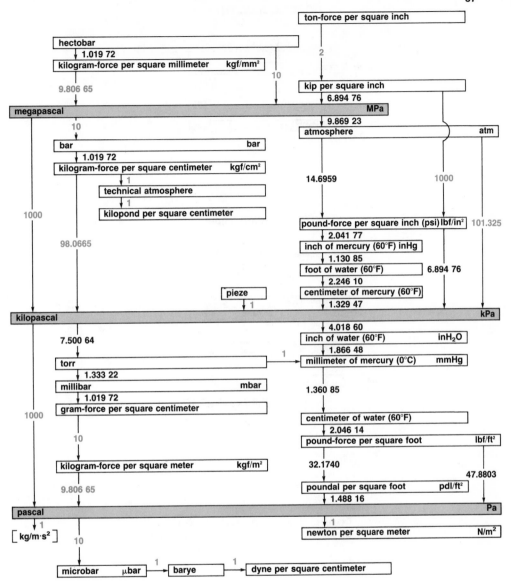

Example: Convert 15 inches of water(60°F) to kilopascals

Solution: 15 inH₂0(60°F) = 15 (÷ 4.018 60) kPa = 3.732 64 kPa

RADIOLOGY

DOSE EQUIVALENT

ACTIVITY (of radionuclides)

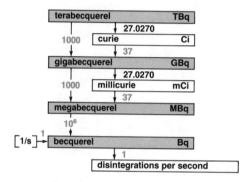

Example: Convert 5 millicuries to gigabecquerels

Solution: 5 mCi = 5 (÷ 27.0270) GBq = 0.185 GBq

RADIOLOGY

EXPOSURE

ABSORBED DOSE

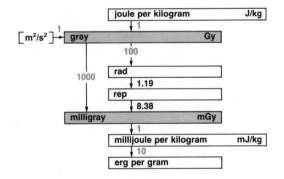

Example: Express 200 rep in terms of joules per kilogram

Solution: 200 rep = 200 (× 8.38) (÷ 1000) (÷ 1) J/kg = 1.68 J/kg

TEMPERATURE

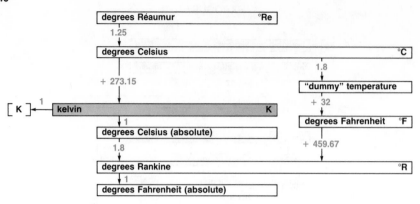

Note: Celsius was formerly called centigrade

The «dummy» temperature is simply a convenient interface between °F and °C

Examples of temperature conversion:

80 degrees Réaumur   = (80 × 1.25)°C = 100°C

100 degrees Celsius   = (100 + 273.15) K = 373.15 K

100 degrees Celsius   = (100 × 1.8) + 32°F = 212°F

108 degrees Fahrenheit = (108 + 459.67) ÷ 1.8 K = (567.67 ÷ 1.8) K = 315.372 K

TEMPERATURE INTERVAL

Note: Btu and calorie are *thermochemical* units

Example: Convert 500 Btu/(h·ft·°F) into watts per meter kelvin
Solution: 500 Btu/(h·ft·°F) = 500 (× 1.729 58) W/(m·K) = 864.79 W/(m·K)

Example: Convert 12000 lbf·ft to kilonewton meters

Solution: 12000 lbf·ft = 12000 (× 1.355 82) (÷ 1000) kN·m = 16.2698 kN·m

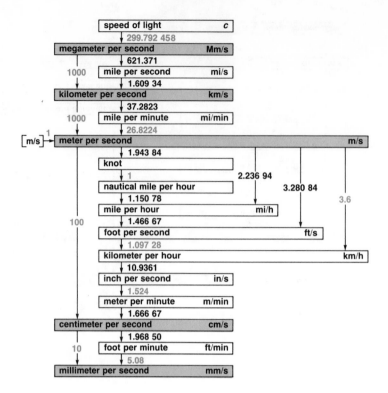

Example: Convert 75 m/s into miles per hour

Solution: 75 m/s = 75 (× 2.236 94) mi/h = 167.77 mi/h

VISCOSITY

DYNAMIC VISCOSITY

KINEMATIC VISCOSITY

VOLUME (including liquid capacity)

VOLUME (including dry capacity)

Example: Convert 5000 cubic feet to cubic meters

Solution: 5000 ft³ = 5000 (÷ 128) (× 3.624 56)m³ = 141.584 m³

APPENDIX I

SELECTED PHYSICAL CONSTANTS [**]

Quantity	Quoted Value	S[*]	SI unit	Symbol
Atomic mass constant	$1.660\ 5402 \times 10^{-27}$	10	kg	m_u
Avogadro constant	$6.022\ 1367 \times 10^{23}$	36	mol^{-1}	N_A
Bohr magneton	$9.274\ 0154 \times 10^{-24}$	31	$J \cdot T^{-1}$	μ_B
Boltzmann constant	$1.380\ 658 \times 10^{-23}$	12	$J \cdot K^{-1}$	$k(=R/N_A)$
Electron charge	$1.602\ 177\ 33 \times 10^{-19}$	49	C	$-e$
Electron specific charge	$-1.758\ 819\ 62 \times 10^{11}$	53	$C \cdot kg^{-1}$	$-e/m_e$
Electron rest mass	$9.109\ 3897 \times 10^{-31}$	54	kg	m_e
Faraday constant	$9.648\ 5309 \times 10^4$	29	$C \cdot mol^{-1}$	F
Fine-structure constant	$0.007\ 297\ 353\ 08$	33	—	α
Josephson frequency to voltage ratio	$4.835\ 9767 \times 10^{14}$	14	$Hz \cdot V^{-1}$	$2e/h$
Magnetic flux quantum	$2.067\ 834\ 61 \times 10^{-15}$	61	Wb	ϕ_o
Molar gas constant	$8.314\ 510$	70	$J \cdot mol^{-1} \cdot K^{-1}$	R
Newtonian gravitational constant	$6.672\ 59 \times 10^{-11}$	85	$m^3 \cdot kg^{-1} \cdot s^{-2}$	G
Permeability of vacuum	$4\pi \times 10^{-7}$	d	H/m	μ_o
Permittivity of vacuum	$8.854\ 1878.. \times 10^{-12}$	d	F/m	ϵ_o
Planck constant	$6.626\ 0755 \times 10^{-34}$	40	$J \cdot s$	h
Planck constant/2π	$1.054\ 572\ 66 \times 10^{-34}$	63	$J \cdot s$	$\hbar(=h/2\pi)$
Quantum of circulation	$3.636\ 948\ 07 \times 10^{-4}$	33	$J \cdot s \cdot kg^{-1}$	$h/2m_e$
Rydberg constant	$1.097\ 373\ 1534 \times 10^7$	13	m^{-1}	R_∞
Standard volume of ideal gas	$22.414\ 10 \times 10^{-3}$	19	$m^3 \cdot mol^{-1}$	V_m
Stefan-Boltzmann constant	$5.670\ 51 \times 10^{-8}$	19	$W \cdot K^{-4} \cdot m^{-2}$	σ
Speed of light	$299\ 792\ 458$	d	$m \cdot s^{-1}$	c

[*] The numbers in this column are the one-standard-deviation uncertainties in the last digits of the quoted value. The symbol d stands here for defined value.

[**] Drawn from E.R. Cohen and B.N. Taylor "The 1986 Adjustment of the Fundamental Physical Constants", Committee on Data for Science and Technology (CODATA) Bulletin Number 63, November 1986.

APPENDIX II

THE INTERNATIONAL SYSTEM OF UNITS

In 1960, at the Eleventh General Conference of Weights and Measures held at Sèvres, France, a world system of units was established, officially named "Système international d'unités". The International System of Units is universally designated by the abbreviation SI.

Base and Derived Units of the SI

The foundation of the International System of Units rests upon seven base units, listed in Table A1.

From these base units other units are derived to express quantities such as area, power, force, magnetic flux, and so forth. Some of these derived units occur so frequently that they have been given special names. A few of these special names are listed in Table A2.

Definitions of SI Base Units

The **meter** (m) is the length of the path travelled by light in vacuum during a time interval of 1/299 792 458 of a second.

The **kilogram** (kg) is the unit of mass; it is equal to the mass of the international prototype of the kilogram. *(The international prototype, made of platinum-iridium is preserved in a vault at Sèvres, France, by the International Bureau of Weights and Measures under conditions specified by the 1st CGPM in 1889).*

The **second** (s) is the duration of 9 192 631 770 periods of the radiation corresponding to the transition between the two hyperfine levels of the ground state of the cesium-133 atom.

TABLE A1	BASE UNITS OF THE SI	
Quantity	**Unit**	**Symbol**
	Base Units	
Length	meter	m
Mass	kilogram	kg
Time	second	s
Electric current	ampere	A
Temperature	kelvin	K
Luminous intensity	candela	cd
Amount of substance	mole	mol

The **ampere** (A) is that constant current which, if maintained in two straight parallel conductors of infinite length, of negligible circular cross section, and placed 1 meter apart in a vacuum, would produce between these conductors a force equal to 2×10^{-7} newton per meter of length.

**TABLE A2   DERIVED UNITS AND SUPPLEMENTARY UNITS
WITH SPECIAL NAMES**

Quantity	Unit	Symbol	Formula
Absorbed dose	gray	Gy	J/kg
Activity (radionuclide)	becquerel	Bq	s^{-1}
Capacitance	farad	F	C/V
Celsius temperature	degree Celsius	°C	K
Dose equivalent	sievert	Sv	J/kg
Electric charge	coulomb	C	A·s
Electric conductance	siemens	S	$1/\Omega$
Electric potential	volt	V	W/A
Electric resistance	ohm	Ω	V/A
Energy	joule	J	N·m
Force	newton	N	$kg·m/s^2$
Frequency	hertz	Hz	1/s
Illumination	lux	lx	$cd·sr/m^2$
Inductance	henry	H	Wb/A
Luminous flux	lumen	lm	cd·sr
Magnetic flux	weber	Wb	V·s
Magnetic flux density	tesla	T	Wb/m^2
Power	watt	W	J/s
Pressure	pascal	Pa	N/m^2
Supplementary Units*			
Plane angle	radian	rad	1
Solid angle	steradian	sr	1

*   In 1980 the Comité International des Poids et Mesures (CIPM) decided to interpret the class of supplementary units in the International System as a class of dimensionless derived units.

The **kelvin** (K), unit of thermodynamic temperature, is the fraction 1/273.16 of the thermodynamic temperature of the triple point of water.

The **candela** (cd) is the luminous intensity, in a given direction, of a source that emits monochromatic radiation of frequency 540×10^{12} hertz and that has a radiant intensity in that direction of 1/683 watt per steradian.

The **mole** (mol) is the amount of substance of a system which contains as many elementary entities as there are atoms in 0.012 kilogram of carbon 12. *Note:* When the mole is used, the elementary entities must be specified and may be atoms, molecules, ions, electrons, other particles or specified groups of such particles.

Definitions of Supplementary Units

The **radian** (rad) is the angle between two radii of a circle which cut off on the circumference an arc equal in length to the radius.

The **steradian** (sr) is the solid angle which, having its vertex in the center of a sphere, cuts off an area of the surface of the sphere equal to that of a square with sides of length equal to the radius of the sphere.

Definitions of Derived Units

Some of the more important derived units are defined as follows:

The **becquerel** (Bq) is the activity of a radionuclide decaying at the rate of one spontaneous nuclear transition per second.

The **coulomb** (C) is the quantity of electricity transported in 1 second by a current of 1 ampere. *(Hence 1 coulomb = 1 ampere second)*

The **degree Celsius** (°C) is equal to the kelvin and is used in place of the kelvin for expressing Celsius temperature (symbol t) defined by the equation $t = T - T_o$ where T is the thermodynamic temperature and $T_o = 273.15$ K, by definition.

The **farad** (F) is the capacitance of a capacitor between the plates of which there appears a difference of potential of 1 volt when it is charged by a quantity of electricity equal to 1 coulomb. *(Hence 1 farad = 1 coulomb per volt)*

The **gray** (Gy) is the absorbed dose when the energy per unit mass imparted to matter by ionizing radiation is one joule per kilogram. *Note:* The gray is also used for the ionizing radiation quantities: specific energy imparted, kerma, and absorbed dose index.

The **henry** (H) is the inductance of a closed circuit in which an electromotive force of 1 volt is produced when the electric current in the circuit varies uniformly at a rate of 1 ampere per second. *(Hence 1 henry = 1 volt second per ampere)*

The **hertz** (Hz) is the frequency of a periodic phenomenon of which the period is 1 second*.

The **joule** (J) is the work done when the point of application of 1 newton is displaced a distance of 1 meter in the direction of the force. *(Hence 1 joule = 1 newton meter)*

The **lumen** (lm) is the luminous flux emitted in a solid angle of 1 steradian by a uniform point source having an intensity of 1 candela. *(Hence 1 lumen = 1 candela steradian)*

The **lux** (lx) is the unit of illuminance equal to one lumen per square meter.

The **newton** (N) is that force which gives to a mass of 1 kilogram an acceleration of 1 meter per second per second. *(Hence 1 newton = 1 kilogram meter per second squared)*

The **ohm** (Ω) is the electric resistance between two points of a conductor when a constant difference of potential of 1 volt, applied between these two points, produces in this conductor a current of 1 ampere, this conductor not being the source of any electromotive force. *(Hence 1 ohm = 1 volt per ampere)*

The **pascal** (P) is the unit of pressure or stress equal to one newton per square meter.

The **siemens** (S) is the unit of electric conductance equal to one reciprocal ohm. *(The siemens was formerly named the mho.)*

The **sievert** (Sv) is the dose equivalent when the absorbed dose of ionizing radiation multiplied by the dimensionless factors Q (quality factor) and N (product of any other multiplying factors) stipulated by the International Commission on Radiological Protection is one joule per kilogram.

The **tesla** (T) is the unit of magnetic flux density equal to one weber per square meter.

The **volt** (V) is the difference of electric potential between two points of a conducting wire carrying a constant current of 1 ampere, when the power dissipated between these points is equal to 1 watt. *(Hence 1 volt = 1 watt per ampere)*

The **watt** (W) is the power which gives rise to the production of energy at the rate of 1 joule per second. *(Hence 1 watt = 1 joule per second)*

The **weber** (Wb) is the magnetic flux which, linking a circuit of one turn, produces in it an electromotive force of 1 volt as it is reduced to zero at a uniform rate in 1 second. *(Hence 1 weber = 1 volt second)*

---

*   The period is the duration of one cycle of a periodic phenomenon. One cycle is any portion of a periodic phenomenon that contains the minimum information needed to describe the periodic phenomenon.

APPENDIX III

DECIBELS AND NEPERS

Power ratio and amplitude ratio

The power ratio of a given power P and an arbitrary reference power P_o is given by

$$\text{power ratio} = P/P_o$$

where P_o, by definition, is the denominator.

The amplitude ratio of a given amplitude F and an arbitrary reference amplitude F_o is given by

$$\text{amplitude ratio} = F/F_o$$

where F_o, by definition, is in the denominator.

In electrotechnology, the amplitude is usually a voltage or current, while in acoustics it is usually a pressure.

Power ratios and amplitude ratios are clearly dimensionless numbers.

Power level difference

Power level difference (symbol L_P) is a quantity whose magnitude is related logarithmically to the power ratio. The unit of power level difference is the decibel (symbol dB). The magnitude L_P of the power level difference is given by the equation:

$$L_P = \{10 \log (P/P_o)\} \, [\text{decibel}] \tag{1}$$

In this equation the term $\{10 \log (P/P_o)\}$ is the numerical value of the quantity, while [decibel] is the unit.

For example, if the given power P is 12.6 W and the reference power P_o is 1 mW, the power level difference is

$$
\begin{aligned}
L_P &= 10 \log (P/P_o) \text{ dB} \\
&= 10 \log (12.6/0.001) \text{ dB} \\
&= 41 \text{ dB}
\end{aligned}
$$

The magnitude of L_P may be positive, negative or zero, depending upon whether P is greater than, less than, or equal to P_o.

In some countries, the unit of power level difference is the neper (symbol Np). The magnitude of the power level difference is then expressed by

$$L_P = \{0.5 \ln (P/P_o)\} \, [\text{neper}] \tag{2}$$

Note that the numerical value is expressed in terms of the natural logarithm of the power ratio.

From equations (1) and (2) we can readily deduce that the [neper] is larger than the [decibel], the relationship being

1 [Np] $=$ 20/(ln10) [dB]

1 [Np] $=$ 8.685 890 [dB]

Thus, a power level difference of 41 dB is the same as a power level difference of 4.72 Np.

Amplitude level difference

Amplitude level difference (symbol L_F) is a quantity whose magnitude is related logarithmically to the amplitude ratio. The unit of amplitude level difference is also the decibel. However, the magnitude of L_F is given by

$$L_F = \{20 \log (F/F_o)\} \text{ [decibels]} \tag{3}$$

Note that the numerical value in equation (3), is different from that in equation (1) because the factor 10 is replaced by the factor 20.

For example, if the reference voltage across a load is 3 V while the actual voltage is 0.7 V, the amplitude level difference L_E is

$L_E =$ 20 log (E/E_o) [dB]

$=$ 20 log (0.7/3) [dB]

$= -12.64$ dB

Whenever the nature of the amplitude is known (voltage, current, pressure, etc.) the amplitude level difference L_F is expressed in terms of voltage, current, pressure, etc. Thus, in the above example, L_E is called the *voltage* level difference. Similarly, in acoustics, we speak of a *pressure* level difference.

The neper is also used as a unit of amplitude level difference. In this case, the magnitude of the amplitude level difference L_F is given by

$$L_F = \ln (F/F_o) \text{ [neper]} \tag{4}$$

The amplitude relationship between the neper and the decibel remains the same as the power relationship: 1 [Np] $=$ 8.685 890 [dB]

Relationship between L_P and L_F

In comparing equations (1) and (3) one may ask why different factors (10 and 20) were chosen for L_P and L_F. The reason is that in many practical cases, the square of the amplitude of a voltage, current or pressure can be used as a proxy for the power at the point of measurement.

For example, in electrotechnology, for a fixed impedance, the power is proportional to the square of the voltage E. This enables us to express the power level difference L_P in terms of the voltage level difference L_E as follows:

$$L_P = 10 \log (P/P_o)$$
$$= 10 \log (E/E_o)^2$$
$$= 20 \log (E/E_o)$$
$$= L_E$$

Thus, when the impedance is fixed $L_P = L_E$. In other words, the amplitude level difference has the same numerical value as the power level difference.

It is usually much easier to measure a voltage, current or pressure than the actual power, and so the concept of amplitude level difference L_F is useful in the determination of L_P.

Standardized amplitude levels and power levels

In science and engineering, it has been found convenient to establish a few standardized amplitudes and powers and use them as reference levels. For example, the international standard for the sound power radiated by a source has been standardized at 1 picowatt. Thus, $P_o = P$ (ref 1 picowatt) $= 10^{-12}$ W.

Similarly, the reference sound pressure has been standardized at 20 micropascals. These standards enable us to express a given acoustic power or a given sound pressure as an absolute sound power level or sound pressure level.

Example:
Calculate the absolute sound pressure level for a sound having an acoustic pressure of 1.2 Pa.

Solution:
$$L_F = 20 \log F/F \text{ (ref 20 μPa)}$$
$$= 20 \log (1.2/20 \times 10^{-6})$$
$$= 95.56 \text{ dB (ref 20 μPa)}$$

In electrical work, the standardized reference power is 1 mW. Consequently, a power of 10 kW corresponds to an absolute power level of

$$L_P = 10 \log P/P \text{ (ref 1mW)}$$
$$= 10 \log (10^4/10^{-3})$$
$$= 70 \text{ dB (ref 1 mW)}$$

When giving power or pressure levels it is important to state the reference power or pressure that was used in making the calculation.

APPENDIX IV

QUANTITIES AND UNITS

Quantities

A quantity is any physical property that can be measured. We use about 400 quantities to describe and measure the physical world around us. A few of these quantities are listed below.

length	work	area	electromotive force
time	energy	density	entropy
mass	speed	angle	pressure
force	acceleration	volume	momentum
torque	power	luminance	viscosity

A quantity may also be a physical constant, such as the gas constant, the Planck constant and the rest mass of an electron.

Relationship between Quantities

The study of physics and mathematics has shown that all quantities are interdependent and so it is always possible to relate them in some way. The relationship can be established by definition, by geometry, by physical law, or by a combination of these three.

Pressure, for example, is a quantity which is related, by definition, to a quantity *force* divided by a quantity *area*. *Area*, in turn, is a quantity related, by geometry, to the product of two quantities of length. *Force*, on the other hand, is a quantity related, by Newton's second law, to the quantity *mass* times the quantity *acceleration*.

Even a seemingly isolated quantity such as *temperature* is related to the quantities *pressure, volume* and *mass* by virtue of the behaviour of gases. We can even relate the quantities *length* and *time* by using that universal constant, the speed of light. Thus, if we define our concepts correctly we can relate any quantity to any other by one or more of the three ways mentioned above.

The relationships between quantities are expressed in the form of *quantity equations*. Thus, the equation *area = length × width* is a quantity equation which states that the quantity (area of a rectangle) is equal to the quantity (length) times the quantity (width).

Base Quantities

To deduce a set of quantity equations we must first establish a number of so-called base quantities. Base quantities are the building blocks upon which we erect the entire structure and relationships of the physical world. The number of base quantities, as well as their choice, is quite arbitrary but, in general, we try to select quantities that are easy to understand, that are frequently used and for which accurate, measurable standards can be set.

In this context, the International System of Units, or SI, makes use of seven base quantities: *mass, length, time, temperature, electric current, luminous intensity* and *amount of substance*.

Derived Quantities

Derived quantities are those which can be deduced by definition, by geometry or by physical law, using the selected base quantities as building blocks. Examples of derived quantities are *velocity* (length/time), *area* (product of two lengths) and *force* (mass × acceleration).

Units

A unit is a quantity that has a definite, defined magnitude. For example, the unit of time we call second is a quantity whose magnitude is defined as follows: "One second is the duration of 9 192 631 770 periods of radiation corresponding to the transition between the two hyperfine levels of the ground state of the cesium - 133 atom".

A unit, therefore, is not just a convenient name, but represents a quantity (time, force, viscosity, etc.) having a precise, stated magnitude. Because units have a definite size, they can be used as measuring sticks to specify the magnitude of a quantity. Clearly, the unit must be of the same nature as the quantity being measured.

As measurement systems evolved, different units were devised, — often to satisfy the needs of a particular specialty in science or engineering. For example, units of pressure such as the bar, inch of mercury, atmosphere and pound per square inch were introduced. The International System of Units was especially created to eliminate the need for such a broad variety of units. In effect, the SI establishes a unique set of units to measure the approximately 400 quantities of practical interest.

Owing to the tremendous range of magnitudes that a quantity may have, it is useful to establish multiples and submultiples of the unit used to measure it. Thus, the kilometer is a unit whose length is precisely 1000 times the defined length of the meter. Similarly, the micrometer is a unit whose length is precisely one millionth of the defined length of the meter.

Magnitude of a quantity

One of the important characteristics of a quantity is its magnitude. Thus, a quantity such as area can extend from the cross section of an atomic nucleus to the size of a football field. Similarly, a quantity like energy can range from that of an orbiting electron to the energy contained in a ton of coal.

As stated previously, the magnitude of any quantity is measured by comparing it with the size of some convenient *unit*. The length of a ship, for example, can be stated by comparing it with the length of the [foot]. The energy in a pile of coal can likewise be stated by comparing it with a unit of energy, such as the [Btu].

The brackets [ ] around "foot" draw our attention to the fact that we are referring to the size of a unit of length, and not simply to some convenient name.

However, the length of the ship could be measured equally well by comparing it with the [yard], and the energy in coal could be specified by comparison with the [joule]. Obviously, the unit we use does not affect the magnitude of the quantity we are measuring, but does affect the numerical value $\hat{N}$ defined below.

The magnitude of a quantity is defined by the equation

$$\text{magnitude of a quantity} = \hat{N} \times [U]$$

in which [U] is the unit and $\hat{N}$ is the number of times the unit must be taken to make up the quantity. $\hat{N}$ is therefore a pure number.

Thus, when we state that a ship is 180 feet long, we are saying, in effect, that

$$\text{magnitude of the length} = 180 \times [\text{foot}]$$

Similarly, when an electric motor develops 400 horsepower we mean

$$\text{magnitude of the power} = 400 \times [\text{horsepower}]$$

which is to say the motor produces 400 times as much power as that of a [horsepower] unit.

These simple statements concerning units and quantities will lead us to some useful conclusions when dealing with equations.

Base Units

For every base quantity in a measurement system, there exists a corresponding base unit. Thus, in the SI, the *kilogram* (kg), *meter* (m), *second* (s), *kelvin* (K), *ampere* (A), *mole* (mol) and *candela* (cd) are the base units (and their symbols) which correspond respectively to the base quantities mass, length, time, temperature, electric current, amount of substance and luminous intensity.

Because base units are the fundamental measuring sticks, it is essential that their magnitude be precisely defined. These definitions are given in Appendix II.

Derived Units

Every derived quantity has a corresponding derived unit. Some of these units are expressed simply as products or as ratios of the base units of which they are composed. This is the case for the derived unit of area (m^2) and for the derived unit of velocity (m/s).

However, it is sometimes useful to give a special name to a derived unit, particularly when its expression in terms of base units is cumbersome. For example, the derived unit of energy in the SI is the kilogram meter squared per second squared ($kg \cdot m^2/s^2$); but it has been given the much shorter name, the joule, (J).

Dimension of Derived Units

Derived units are said to possess a *dimension*. The dimension of a derived unit is its value in terms of base units. For example, the base dimension of the [newton] (symbol N) is [$kg \cdot m/s^2$]. Thus, $1 \text{ N} = 1 \text{ } kg \cdot m/s^2$.

The SI dimensions of the units of force, power, voltage, magnetic flux, etc., are displayed between brackets on the left-hand side of each conversion chart. Some have surprising dimensions, leading us to think we may have discovered the very essence of truth. The volt, for example, has the dimension [$kg \cdot m^2/A \cdot s^3$] which means exactly what it says – one volt is equal to one kilogram meter squared per ampere second cubed. But lest we attach some deep meaning to this dimension, we should remember that it all depends upon the quantity equations, and how the base units were defined.

Standards

All base units are essentially *definitions* of physical quantities, and so they are always exact. Base units are measured in the real world by *standards* which are the physical embodiment of the base units. These standards can approach the defined values of the base units very closely, but because the standards are affected by temperature, pressure and other extraneous variables, they measure the base unit with high accuracy only under prescribed conditions. Typical accuracies that are attainable in standards laboratories for SI base units are listed below.

kilogram: 1 part in 10^8	ampere: 1 part in 10^6
meter: 1 part in 10^{10}	kelvin: 1 part in 10^6
second: 1 part in 10^{13}	candela: 5 parts in 10^3
	mole: 1 part in 10^6

APPENDIX V

QUANTITY EQUATIONS AND
NUMERICAL EQUATIONS

In dealing with scientific relationships it is important to distinguish between two types of equations: quantity equations and numerical equations. Both types are encountered in texts and reference books and the concept of units and quantities is useful in understanding their respective attributes.

Quantity Equations

The energy in a hurricane, the pressure at the bottom of the sea, the weight of a stone and the viscosity of oil are quantities of nature which existed long before man was on earth to measure them. And, whether man measures them or not, these quantities are there, interacting with each other according to fundamental laws. Because quantities — and not units — conform to the laws of nature, physicists often express these laws in terms of *quantity equations.**

Quantity equations possess two important attributes:

      1.   They show the relationship between quantities.

      2.   They can be used with any system of units.

There are three basic types of quantity equations:

1.   Quantity equations established from the laws of nature; example:

    Newton's second law of motion

$$F = ma$$

    in which

        F = magnitude of the force
       m = magnitude of the mass
       a = magnitude of the acceleration

2.   Quantity equations established from a definition; example:

    definition of pressure

$$p = F/A$$

---

* quantity equations are also called physical equations, or equations between quantities.

in which

p = magnitude of the pressure
F = magnitude of the force
A = magnitude of the area

3. Quantity equations arising from geometry; example:

Area of a circle

$$A = \pi r^2$$

in which

A = magnitude of the area
r = magnitude of the radius
π = coefficient based upon the geometry of a circle.

None of these equations imposes a particular set of units. Consequently, we are free to use any convenient units to describe the magnitudes of the quantities F, m, a, p, A and r.

Many quantity equations are a combination of the three basic types mentioned above, but in all cases we can use any units we please to describe the magnitudes of the quantities that are involved.

Some quantity equations contain a physical constant, such as the constant R_a in the equation giving the properties of dry air:

$$pV = R_a mT$$

in which

p = magnitude of the pressure
V = magnitude of the volume
m = magnitude of the mass (of dry air)
T = magnitude of the absolute temperature
R_a = magnitude of the constant for dry air = 287 J/(kg·K)

The magnitude of R_a is expressed in SI units, but this places no restriction whatever on the units that may be used for the other quantities p, V, m and T.

The freedom to use any system or *combination* of units is the reason why quantity equations are so useful. Their elegant simplicity also reflects the basically simple laws of nature and the orderly structure of geometry. The following example illustrates the versatility of these equations in accommodating any set of units.

Example:
Calculate the value of the earth's centrifugal force which acts on a 150-pound man located at the equator. The radius of the earth is 4000 miles and its angular velocity is 1 revolution/day.

Solution:
The quantity equation for centrifugal force is

$$F = m\omega^2 r$$

in which F is the force, m the mass, ω the angular velocity and r the radius (expressed in any units at all).

$$
\begin{aligned}
F &= m\omega^2 r \\
&= 150 \text{ lb} \times (1 \text{ r/d})^2 \times 4000 \text{ mi} \\
&= 600\,000 \ \frac{\text{lb·r}^2\text{·mi}}{\text{d}^2}
\end{aligned}
$$

(r = revolution ; d = day)

Referring to the appropriate charts (page 30, 9, 25, 42, 24) in order to convert all units to SI base units we find:

$$F = 600\,000 \ \frac{\text{kg}}{2.204\,62} \times \frac{(4 \times 1.570\,80)^2}{(24 \times 3600 \text{ s})^2} \times (1.609\,344 \times 1000) \text{ m}$$

$$
\begin{aligned}
&= 2.315 \text{ kg·m/s}^2 \\
&= 2.315 \text{ N}
\end{aligned}
$$

The centrifugal force is 2.315 N, or 0.520 lbf (pound-force).

Numerical Equations

We have seen that quantity equations can be used with any system of units. However, it is sometimes convenient to set up an equation to accommodate a *particular* set of units. Such an equation is called a *numerical equation** because all its algebraic terms stand for pure numbers.

We can readily transform a quantity equation into a numerical equation (1) by specifying the units we wish to employ, and (2) by using the following procedure.

1. The quantities Q_1, Q_2, ... Q_n in the quantity equation are respectively replaced by the terms $\hat{N}_1[U_1]$, $\hat{N}_2[U_2]$, ... $\hat{N}_n[U_n]$, where $[U_1]$, $[U_2]$ ... $[U_n]$ are the respective units we wish to employ, and the symbols $\hat{N}_1$, $\hat{N}_2$ ... $\hat{N}_n$ represent pure numbers.

---

* numerical equations are also called equations between numerical values.

2. The resulting equation is simplified by grouping the units $[U_1, [U_2] \ldots [U_n]$ together. Each unit is then expressed in terms of SI base units. It will be found that the base units all cancel, leaving us with a pure number $\hat{k}$ whose value depends upon the units $[U_1], [U_2], \ldots [U_n]$ that were originally selected.

3. The final numerical equation is therefore composed of the symbols $\hat{N}_1$, $\hat{N}_2$, $\ldots \hat{N}_n$ (each of which represents a pure number) and the pure number $\hat{k}$.

In referring to technical literature, how can we tell the difference between a quantity equation and a numerical equation? In a numerical equation, each algebraic term represents the *numerical* value of the quantity and specifies the unit that applies thereto. On the other hand, in a quantity equation, each algebraic term represents only the magnitude of each quantity and no units are specified. The following example illustrates the difference between the two, using the equation for the centrifugal force that acts on a revolving body.

Quantity Equation for centrifugal force

$$F = m\omega^2 r$$

in which

$F = $ centrifugal force

$m = $ mass

$\omega = $ angular velocity

$r = $ length of radius

Numerical Equation for centrifugal force

$$\hat{F} = 0.003342 \ \hat{m}\hat{\omega}^2\hat{r}$$

in which

$\hat{F} = $ centrifugal force, in newtons

$\hat{m} = $ mass, in kilograms

$\hat{\omega} = $ angular velocity, in revolutions per minute

$\hat{r} = $ length of radius, in feet

$0.003342 = $ numerical coefficient

In the above quantity equation any units may be used, but in the numerical equation the mass must be in kilograms, the angular velocity must be in revolutions per minute and the radius must be in feet. Furthermore, the resulting centrifugal force is expressed in newtons.

How is the numerical equation arrived at? It is derived from the quantity equation using the procedure described above. The detailed methodology will now be given, knowing that we want to express F in newtons, m, in kilograms, ω in revolutions per minute and r in feet. (In order to clearly distinguish between algebraic symbols that represent quantities from those that represent numerical values, we have put a "hat" on the latter).

First, we have the quantity equation

$$F = m\omega^2 r$$

but

$$F = \text{(number of force units)} \times \text{[desired force unit]}$$
$$= \hat{N}_1[U_1]$$
$$= \hat{F}\ [N]$$

and

$$m = \text{(number of mass units)} \times \text{[desired mass unit]}$$
$$= \hat{N}_2[U_2]$$
$$= \hat{m}\ [kg]$$
$$\omega = \text{(number of angular velocity units)} \times \text{[desired angular velocity unit]}$$
$$= \hat{N}_3[U_3]$$
$$= \hat{\omega}\ [\text{revolution per minute}]$$
$$= \hat{\omega}\ [r/min]$$
$$r = \text{(number of length units)} \times \text{[desired length unit]}$$
$$= \hat{N}_4[U_4]$$
$$= \hat{r}\ [ft]$$

Substituting the magnitude of these quantities in the quantity equation, we obtain

$$F = m\omega^2 r$$
$$\hat{F}\ [N] = \hat{m}\ [kg]\ (\hat{\omega}\ [r/min])^2\ \hat{r}\ [ft]$$

grouping the units together, we obtain

$$\hat{F} = \hat{m}\hat{\omega}^2\hat{r}\ [kg]\ [r/min]^2\ [ft]\ [N^{-1}]$$

then, expressing the units in terms of SI base units

$$\hat{F} = \hat{m}\hat{\omega}^2\hat{r}\ [kg]\ [2\pi/60s]^2\ [m/(1.093\ 61 \times 3)]\ [N^{-1}]$$

$$= \hat{m}\hat{\omega}^2\hat{r}\ \frac{(2\pi/60)^2}{3.281}[kg{\cdot}m/s^2]\ [N^{-1}]$$

Finally, recalling that $1\ N = kg{\cdot}m/s^2$, we see that all the base units cancel and so we obtain the resulting numerical equation:

$$\hat{F} = 0.003342\ \hat{m}\hat{w}^2\hat{r}$$

The above derivation shows clearly that the algebraic symbols $\hat{F}$, $\hat{m}$, $\hat{\omega}$, $\hat{r}$ represent numerical values and the coefficient $\hat{k} = 0.003342$ is also a pure number.

It is obvious that we can generate as many numerical equations as we please from a given quantity equation. The number is limited only by the variety of units we wish to employ. Furthermore, whenever the units are changed, the value of the coefficient k changes.

Meaning of Coherence

A system of units is said to be coherent when the numerical equations, expressed in base units, have the same numerical factors as the corresponding quantity equations. Coherence is one of the distinguishing features of the SI. The following list of equations illustrates the meaning of coherence.

Quantity Equation	Numerical Equation
$W = Fd$	$\hat{W} = \hat{F}\hat{d}$
$F = ma$	$\hat{F} = \hat{m}\hat{a}$
$E = 1/2\ mv^2$	$\hat{E} = 1/2\ \hat{m}\hat{v}^2$
$F = G\ m_1 m_2/d^2$	$\hat{F} = 6.673 \times 10^{-11}\ \hat{m}_1\hat{m}_2/\hat{d}^2$

in which

in which

$W = $ work	$\hat{W} = $ work in joules
$F = $ force	$\hat{F} = $ force in newtons
$d = $ distance	$\hat{d} = $ distance in meters
$E = $ energy	$\hat{E} = $ energy in joules
$m = $ mass	$\hat{m} = $ mass in kilograms
$v = $ velocity	$\hat{v} = $ velocity in meters per second
$a = $ acceleration	$\hat{a} = $ acceleration in meters per second squared
$G = 6.673 \times 10^{-11}\ \text{m}^3{\cdot}\text{kg}^{-1}{\cdot}\text{s}^{-2}.$	

Note that the coefficient G is composed of a numerical factor and base units.

Note that the numerical factor 6.673×10^{-11} is the same as in the quantity equation.

Converting a numerical equation into a quantity equation

A numerical equation can readily be converted into a quantity equation, thus rendering it universal as far as units are concerned. The procedure, described below, is the reverse of that used to convert a quantity equation into a numerical equation:

1. The terms $\hat{N}_1, \hat{N}_2 \ldots \hat{N}_n$ in the numerical equation are replaced by the terms $Q_1/[U_1], Q_2/[U_2], \ldots Q_n/[U_n]$, where $Q_1, Q_2, \ldots Q_n$ represent the respec-

tive quantities, and $[U_1]$, $[U_2]$ ... $[U_n]$ are the corresponding units specified in the numerical equation.

2. The resulting equation is then simplified by grouping the units $[U_1]$, $[U_2]$ ... $[U_n]$ together, and expressing each in terms of SI base units. The units are manipulated as if they were algebraic symbols. Consequently, they can be reduced to their most simple algebraic expression whenever this is convenient.

3. The resulting quantity equation is composed of the symbols Q_1, Q_2, ... Q_n and a coefficient K consisting of a pure number and one or more base units. In many important cases, the coefficient has the dimensionless value of 1.

As an example of such a conversion, consider the following numerical equation giving the fusing current of a copper conductor in relation to its diameter.

$$\hat{I} = 70 \, \hat{d}^{1.5}$$

in which

$\hat{I} =$ current, in amperes (in this case, $\hat{I}$ corresponds to $\hat{N}_1$)
$\hat{d} =$ diameter, in millimeters (in this case, $\hat{d}$ corresponds to $\hat{N}_2$)

This equation is typical of many empirical equations in which the relationships are determined experimentally.

First, we know that,

magnitude of current $= I =$ (number of current units) $\times$ [given current unit]

$$= \hat{N}_1[U_1]$$
$$= \hat{I} \, [A]$$

therefore

$$\hat{I} = I/[A]$$

Furthermore,

magnitude of diameter $= d =$ (number of length units) $\times$ [given length unit]

$$= \hat{d} \, [mm]$$

therefore

$$\hat{d} = d/[mm]$$

Referring to the numerical equation, if we now substitute the numerical values $\hat{I}$ and $\hat{d}$ by the respective magnitudes divided by the corresponding unit, we obtain

$$\hat{I} = 70\ \hat{d}^{1.5}$$

$$
\begin{aligned}
I/[A] &= 70\ (d/[\text{mm}])^{1.5} \\
I &= d^{1.5}\ 70\ [A]\ [\text{mm}]^{-1.5} \\
&= d^{1.5}\ 70\ [A]\ [\text{m}/1000]^{-1.5} \\
&= d^{1.5}\ 70\ [A]\ [\text{m}]^{-1.5}\ [1000]^{1.5} \\
&= 2.21 \times 10^{6}\ [A]\ [\text{m}]^{-1.5}\ d^{1.5}
\end{aligned}
$$

Thus, we obtain the quantity equation

$$I = k\ d^{1.5}$$

in which

I = magnitude of the current

d = magnitude of the diameter

k = constant whose magnitude is 2.21×10^{6} A·m$^{-1.5}$

This example also shows that units can be treated like algebraic symbols, as far as exponents are concerned.

Discovering Dimensions

To determine the dimension of a derived unit we must know two things:

1. The definition of the derived unit,

2. The quantity equation which connects the derived unit with units whose dimensions are already known.

Let us determine, for example, the dimension of the newton. The SI definition states that "the newton is that force which gives to a mass of one kilogram an acceleration of one meter per second per second".

The quantity equation is

$$F = ma$$

Based upon the definition, we can write:

$$
\begin{aligned}
\text{one [newton]} &= \text{one[kilogram]} \times \text{one [meter per second per second]} \\
1\ [N] &= 1\ [\text{kg}] \times 1\ [\text{m/s}^2] \\
\text{whence N} &= \text{kg·m·s}^{-2}
\end{aligned}
$$

Knowing the dimension of the newton, we can now determine the dimension of the joule. By definition, "the joule is the work done when the point of application of one newton is displaced a distance of one meter in the direction of the force".

The quantity equation is

$$E = Fd$$

that is,   work = force $\times$ distance

Using the definition and the quantity equation, we can therefore write:

$$\text{one [joule]} = \text{one [newton]} \times \text{one [meter]}$$
$$1 \text{ [J]} = 1 \text{ [N]} \times 1 \text{ [m]}$$

thus

$$J = N \cdot m$$

but since

$$N = kg \cdot m \cdot s^{-2}$$

we obtain

$$J = kg \cdot m^2 \cdot s^{-2}$$

and so the dimension of the joule is $kg \cdot m^2 \cdot s^{-2}$

We are now in position to determine the dimension of the watt. By definition, "the watt is the power which gives rise to the production of energy at the rate of one joule per second. The quantity equation is

$$P = E/t$$

that is, power $= \dfrac{\text{energy}}{\text{time}}$

Using the definition, we can therefore write

$$\text{one [watt]} = \frac{\text{one [joule]}}{\text{one [second]}}$$

$$1 \text{ [W]} = \frac{1 \text{ [J]}}{1 \text{ [s]}}$$

therefore $W = J \cdot s^{-1}$

but since $J = kg \cdot m^2 \cdot s^{-2}$

we obtain $W = kg \cdot m^2 \cdot s^{-3}$

and so the dimension of the watt is $kg \cdot m^2 \cdot s^{-3}$.

Let us now discover the dimension of the volt. By definition "one volt is the difference of potential between two points of a conducting wire carrying a constant current of one ampere, when the power dissipated between these points is equal to one watt".

The quantity equation is

$$U = P/I$$

that is,

$$electromotive\ force\ =\ \frac{power}{current}$$

Using the definition in the quantity equation, we can therefore write:

$$one\ [volt]\ =\ \frac{one\ [watt]}{one\ [ampere]}$$

$$1\ [V]\ =\ \frac{1\ [W]}{1\ [A]}$$

therefore $V\ =\ W \cdot A^{-1}$

but since $W\ =\ kg \cdot m^2 \cdot s^{-3}$

we obtain $V\ =\ kg \cdot m^2 \cdot A^{-1} \cdot s^{-3}$

and so the dimension of the volt is $kg \cdot m^2 \cdot A^{-1} \cdot s^{-3}$

The dimensions of derived units are found, therefore, by a pyramiding process in which each successive derived unit becomes the stepping-stone to establish the next. That is how the dimension of each quantity listed in the charts was established. It is, however, important to remember that these dimensions are all derived from the base units of the SI system of measurement.

BIBLIOGRAPHY

Ambler, E. 1971. *SI Units, Philosophical Basis for the Base Units.* Technical News Bulletin, March 1971, National Bureau of Standards, Washington, DC.

ANSI/IEEE, 1982. *American National Standard Metric Practice.* (ANSI/IEEE Std 268-1982) New York : Institute of Electrical and Electronics Engineers.

BIPM, 1985. *The International System of Units (SI).* Sèvres, France. Bureau International des Poids et Mesures.

Cohen, E.R. and Taylor B.N. 1986. *The 1986 Adjustment of the Fundamental Physical Constants.* Codata Bulletin. New York. Pergamon.

CSA, 1979. *Canadian Metric Practice Guide. CAN3-Z234.1-79.* Toronto. Canadian Standards Association.

Drazil, J.V. 1983. *Quantities and Units of Measurement.* London. Mansell Publishing.

Engrand, J.C. 1981. *Units and Their Equivalences.* Paris. Vuibert.

Horvath, A.L. 1987. *Conversion Tables of Units for Science and Engineering.* New York. Elsevier.

Huntley, H.E. 1967. *Dimensional Analysis.* New York. Dover Publications.

International Electrotechnical Commission, 1981. *IEC - Publication 27.* Geneva. IEC.

International Commission of Radiation Quantities and Units, 1980. *Radiation Quantities and Units, ICRU Report 33.* Washington. ICRU.

International Organization for Standardization, 1981. *ISO - 31 Series.* Geneva. ISO.

Ipsen, D.C. 1960. *Units, Dimensions and Dimensionless Numbers.* New York, McGraw-Hill.

Judson, L.V. 1976. *Weights and Measures Standards of the United States: A Brief History.* National Bureau of Standards publication 447.

Kerwin, L. 1978. International Union of Pure and Applied Physics. *Symbols, Units and Nomenclature in Physics.* Québec. Université Laval.

McGlashan, M.L. 1973. *Manual of Symbols and Terminology for Physico-Chemical Quantities and Units.* International Union of Pure and Applied Chemistry (IUPAC). New York, Pergamon.

Martinek, A. *Metric System (SI) in Engineering Technology.* Waterloo. Reeve Bean.

Maxwell, J.C. 1954. *A Treatise on Electricity and Magnetism,* Vol. 2 Chap X. New York. Dover Publications.

Preston-Thomas, H. et al. 1968. *An Absolute Measurement of the Acceleration Due to Gravity*. NRC Publication 5693. Ottawa. National Research Council of Canada.

Wildi, T. 1970. *Units*. Québec. Volta.
————— 1972. *Understanding Units*. Québec. Volta.

Young, L. 1969. *Systems of Units in Electricity and Magnetism*. Edinburgh. Oliver and Boyd.

INDEX